AF821052

THIS BOOK BELONGS TO:

COLOR THE NUMBER, THEN TRACE IT.

ZERO

0

Zero Zero Zero Zero

Zero Zero Zero Zero

Zero Zero Zero Zero

Zero Zero Zero Zero

LET'S TRACE THE NUMBER.

ZERO

0 0 0 0 0 0 0 0

0 0 0 0 0 0 0 0

0 0 0 0 0 0 0 0

0 0 0 0 0 0 0 0

COLOR THE NUMBER, THEN TRACE IT.

ONE

One One One One

One One One One

One One One One

One One One One

LET'S TRACE THE NUMBER.

ONE

1 1 1 1 1 1 1

1 1 1 1 1 1 1

1 1 1 1 1 1 1

1 1 1 1 1 1 1

COLOR THE NUMBER, THEN TRACE IT.

TWO

2

Two Two Two Two

Two Two Two Two

Two Two Two Two

Two Two Two Two

LET'S TRACE THE NUMBER.

TWO

2 2 2 2 2 2 2

2 2 2 2 2 2 2

2 2 2 2 2 2 2

2 2 2 2 2 2 2

COLOR THE NUMBER, THEN TRACE IT.

THREE

Three Three Three

Three Three Three

Three Three Three

Three Three Three

LET'S TRACE THE NUMBER.

THREE

3 3 3 3 3 3 3

3 3 3 3 3 3 3

3 3 3 3 3 3 3

3 3 3 3 3 3 3

COLOR THE NUMBER, THEN TRACE IT.

FOUR

Four Four Four Four

Four Four Four Four

Four Four Four Four

Four Four Four Four

LET'S TRACE THE NUMBER.

FOUR

4 4 4 4 4 4 4

4 4 4 4 4 4 4

4 4 4 4 4 4 4

4 4 4 4 4 4 4

COLOR THE NUMBER, THEN TRACE IT.

FIVE

5

Five Five Five Five

Five Five Five Five

Five Five Five Five

Five Five Five Five

LET'S TRACE THE NUMBER.

FIVE

5 5 5 5 5 5 5

5 5 5 5 5 5 5

5 5 5 5 5 5 5

5 5 5 5 5 5 5

COLOR THE NUMBER, THEN TRACE IT.

SIX

Six Six Six Six

Six Six Six Six

Six Six Six Six

Six Six Six Six

LET'S TRACE THE NUMBER.

SIX

6 6 6 6 6 6 6

6 6 6 6 6 6 6

6 6 6 6 6 6 6

6 6 6 6 6 6 6

COLOR THE NUMBER, THEN TRACE IT.

SEVEN

7

Seven Seven Seven

Seven Seven Seven

Seven Seven Seven

Seven Seven Seven

LET'S TRACE THE NUMBER.

SEVEN

7

COLOR THE NUMBER, THEN TRACE IT.

EIGHT

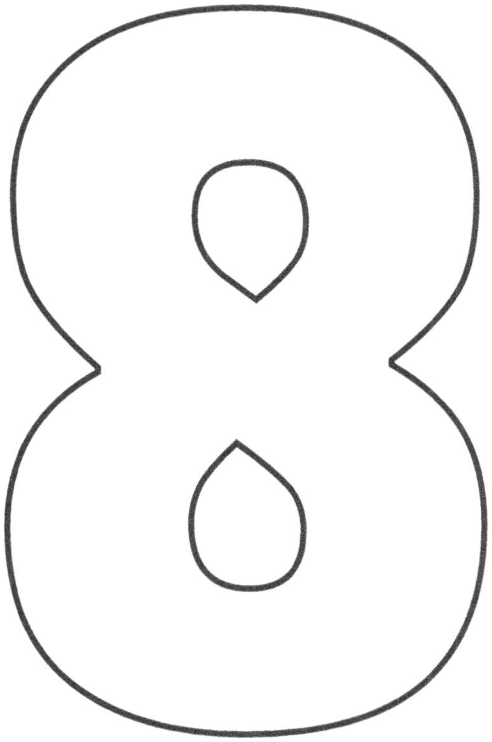

Eight Eight Eight Eight

Eight Eight Eight Eight

Eight Eight Eight Eight

Eight Eight Eight Eight

LET'S TRACE THE NUMBER.

EIGHT

8 8 8 8 8 8 8

8 8 8 8 8 8 8

8 8 8 8 8 8 8

8 8 8 8 8 8 8

COLOR THE NUMBER, THEN TRACE IT.

NINE

9

Nine Nine Nine Nine

Nine Nine Nine Nine

Nine Nine Nine Nine

Nine Nine Nine Nine

LET'S TRACE THE NUMBER.

NINE

9 9 9 9 9 9 9

9 9 9 9 9 9 9

9 9 9 9 9 9 9

9 9 9 9 9 9 9

COLOR THE NUMBER, THEN TRACE IT.

TEN

10

Ten Ten Ten Ten

Ten Ten Ten Ten

Ten Ten Ten Ten

Ten Ten Ten Ten

LET'S TRACE THE NUMBER.

TEN

10 10 10 10 10 10

10 10 10 10 10 10

10 10 10 10 10 10

10 10 10 10 10 10

1 One

COLOR 1 APPLE

2 Two

COLOR THE STRAWBERRY THAT HAS 2 SEEDS

3 Three

COLOR 3 LEMONS

4 Four

COLOR 4 SEA CANCERS

5 Five

COLOR 5 ROCKETS

6 Six

COLOR THE WATERMELON THAT HAS 6 SEEDS

7 Seven

COLOR 7 CUPS

8 Eight

COLOR 8 BALLOONS

9 Nine

COLOR 9 BEES

10 Ten

COLOR THE LADYBUG THAT HAS 10 DOTS

COLORING GAME

READ EACH NUMBER, THEN COLOR AS MANY AS MANY PICTURES AS YOU READ.

1

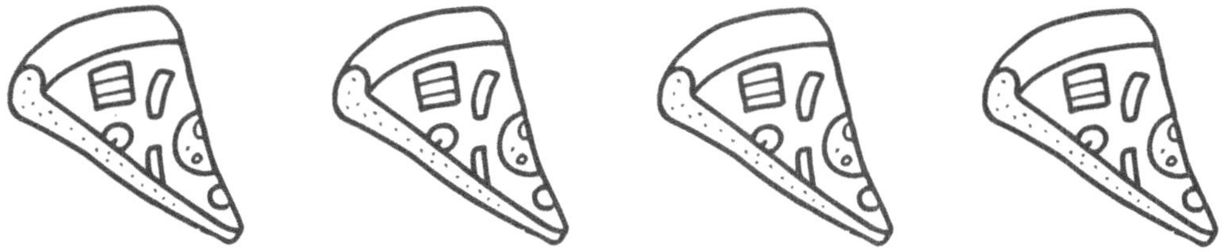

2

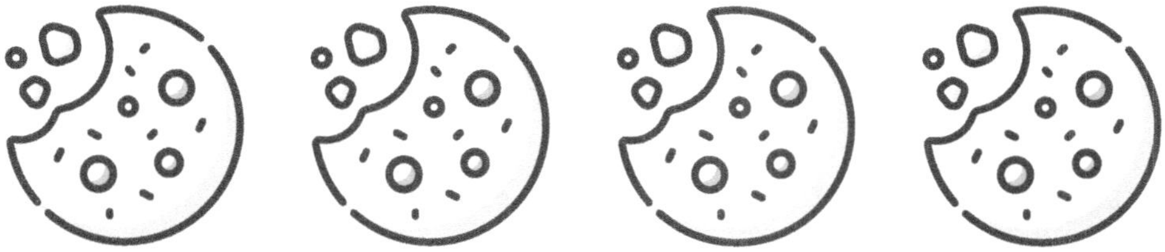

4

COLORING GAME

READ EACH NUMBER, THEN COLOR AS MANY AS MANY PICTURES AS YOU READ.

5

3

6

COLORING GAME

READ EACH NUMBER, THEN COLOR AS MANY AS MANY PICTURES AS YOU READ.

8

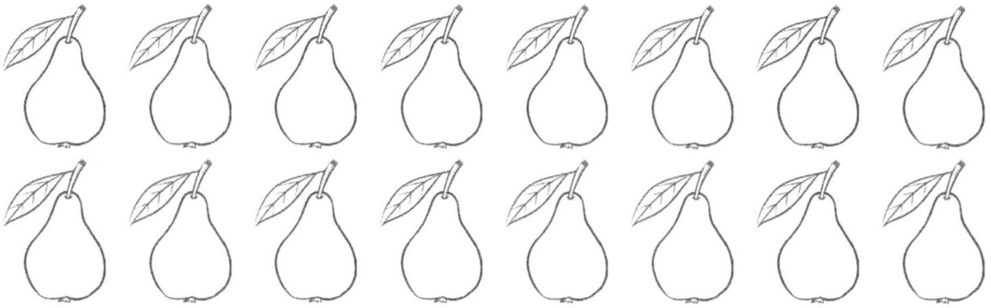

7

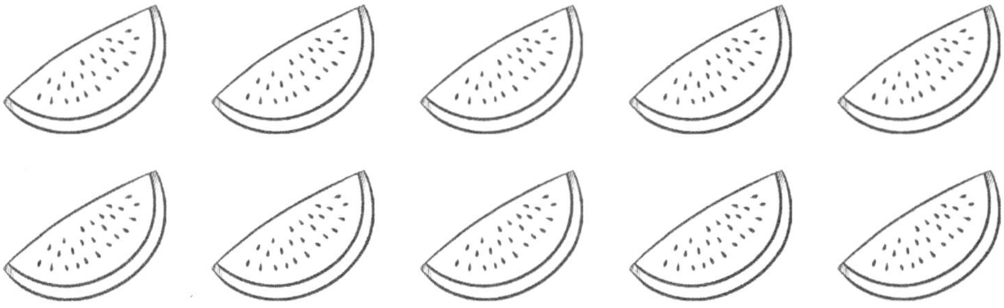

9

COLORING GAME

READ EACH NUMBER, THEN COLOR AS MANY AS MANY PICTURES AS YOU READ.

10

12

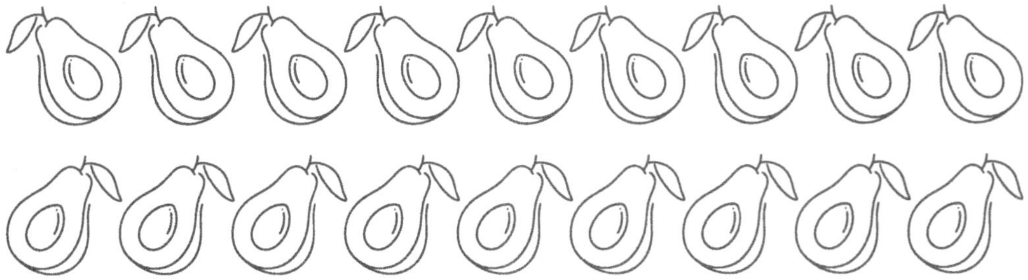

15

COLORING GAME

READ EACH NUMBER, THEN COLOR AS MANY AS MANY PICTURES AS YOU READ.

11

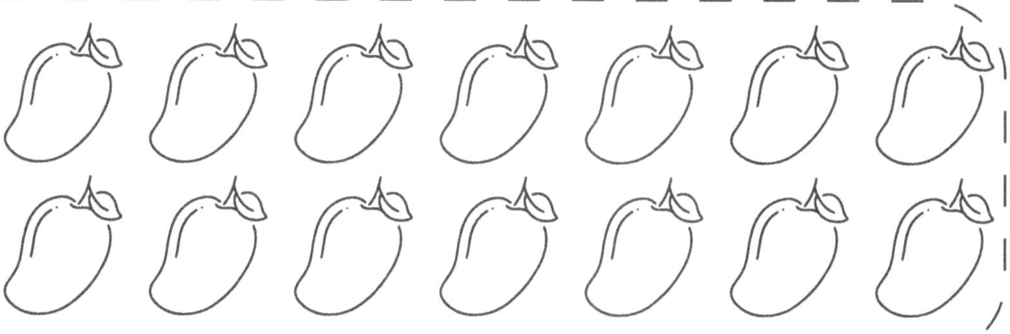

16

20

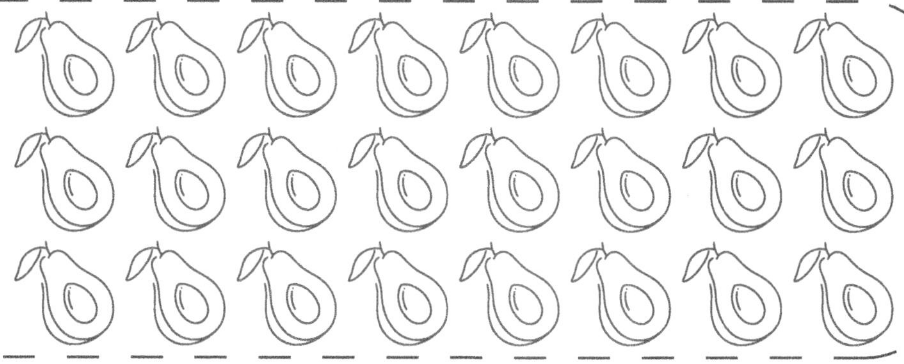

COLORING GAME

READ EACH NUMBER, THEN COLOR AS MANY AS MANY PICTURES AS YOU READ.

17

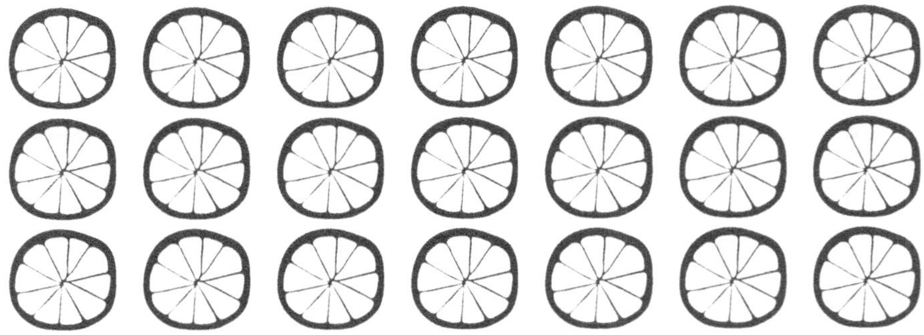

14

10

COLORING GAME

READ EACH NUMBER, THEN COLOR AS MANY AS MANY PICTURES AS YOU READ.

24

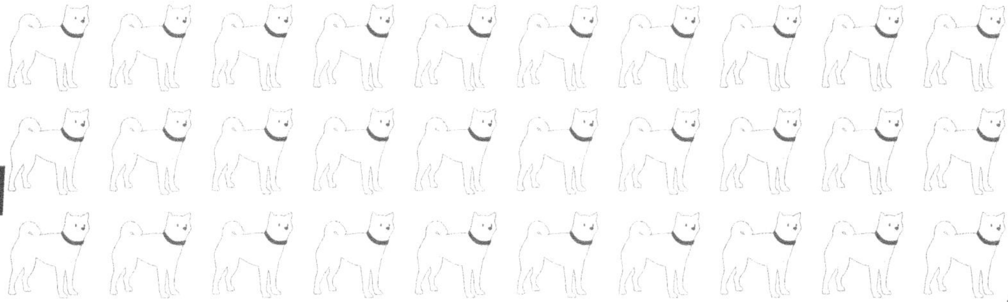

12

11

COLORING GAME

READ EACH NUMBER, THEN COLOR AS MANY AS MANY PICTURES AS YOU READ.

32

27

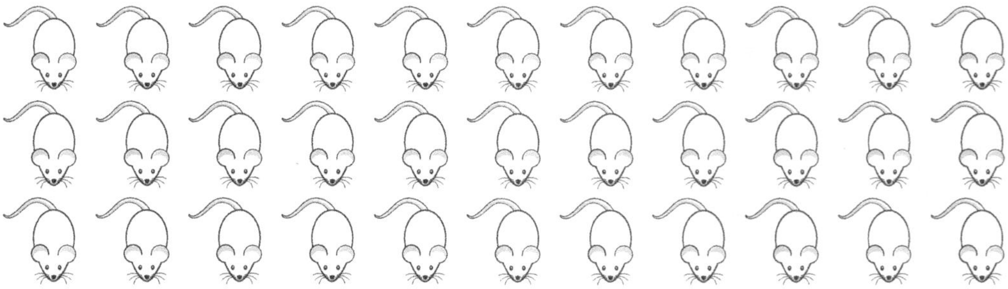

5

COLORING GAME

READ EACH NUMBER, THEN COLOR AS MANY AS MANY PICTURES AS YOU READ.

16

13

36

COLORING GAME

READ EACH NUMBER, THEN COLOR AS MANY AS MANY PICTURES AS YOU READ.

11

9

7

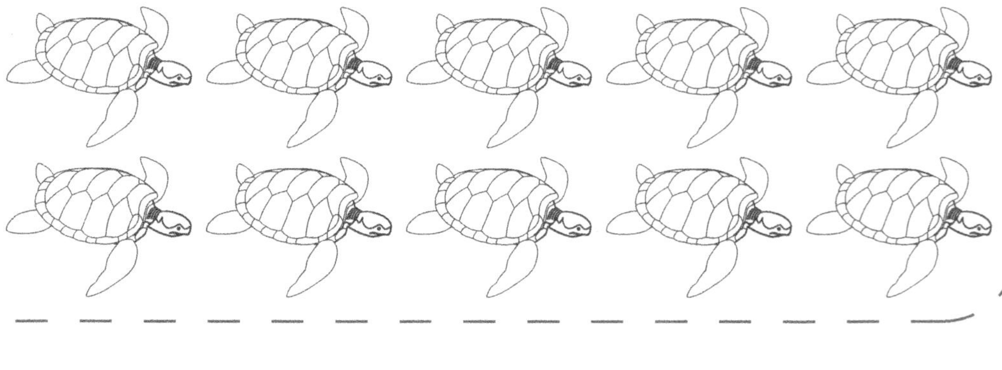

COUNTING GAME

COUNT HOW MANY MANY PICTURES IN THE BOX, THEN WRITE THE NUMBER.

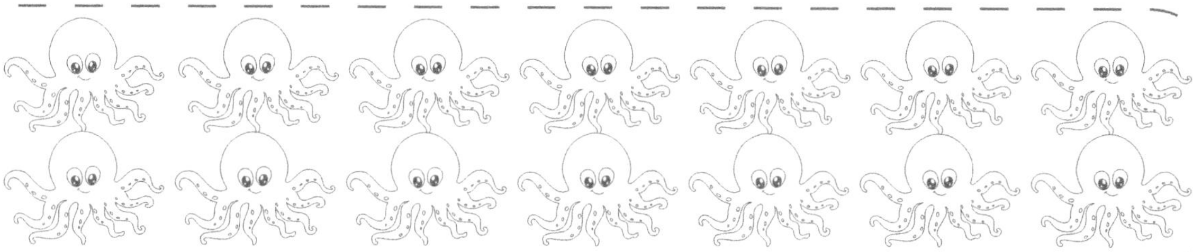

COUNTING GAME

COUNT HOW MANY MANY PICTURES IN THE BOX, THEN WRITE THE NUMBER.

COUNTING GAME

COUNT HOW MANY MANY PICTURES IN THE BOX, THEN WRITE THE NUMBER.

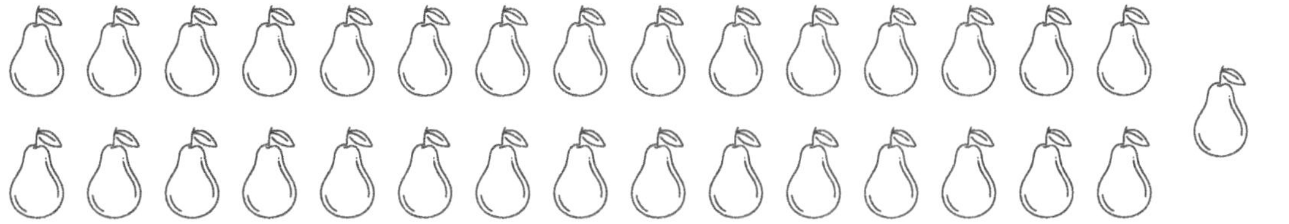

COUNTING GAME

COUNT HOW MANY MANY PICTURES IN THE BOX, THEN WRITE THE NUMBER.

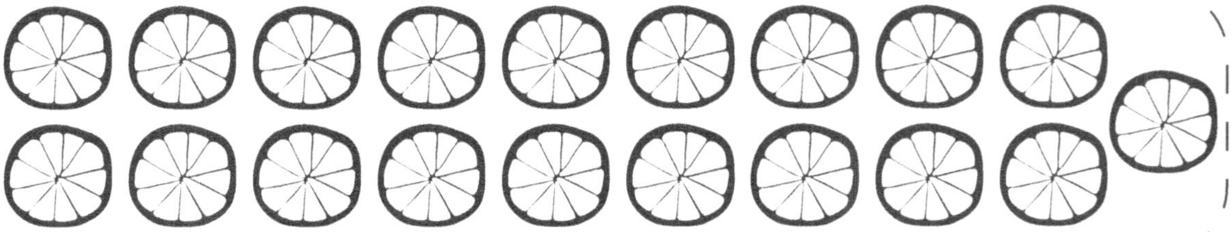

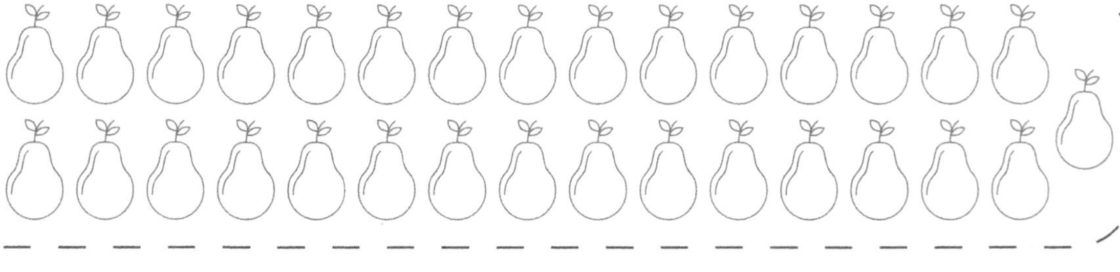

COUNTING GAME

COUNT HOW MANY MANY PICTURES IN THE BOX, THEN WRITE THE NUMBER.

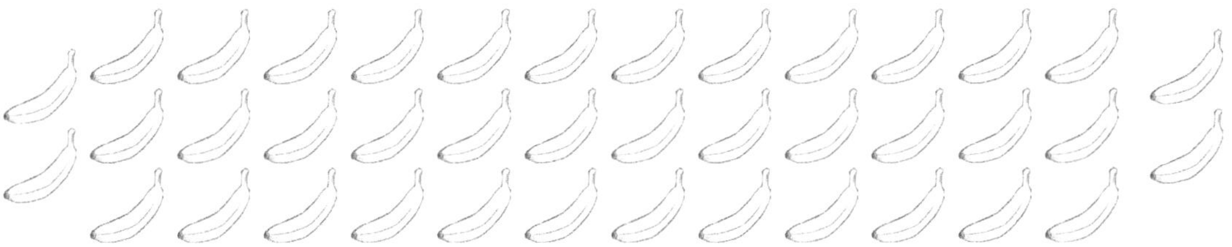

COUNTING GAME

COUNT HOW MANY MANY PICTURES IN THE BOX, THEN WRITE THE NUMBER.

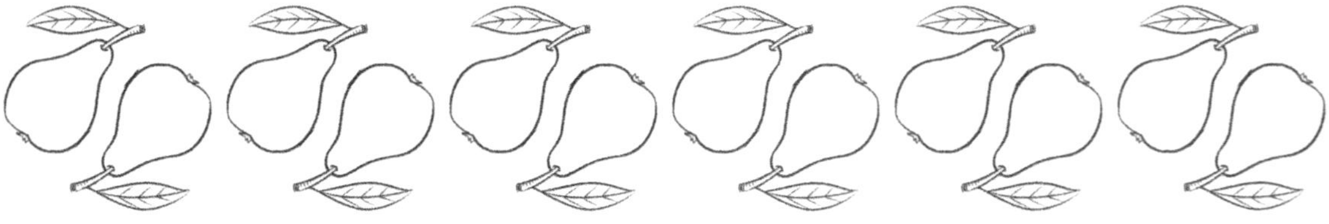

COUNTING GAME

COUNT HOW MANY MANY PICTURES IN THE BOX, THEN WRITE THE NUMBER.

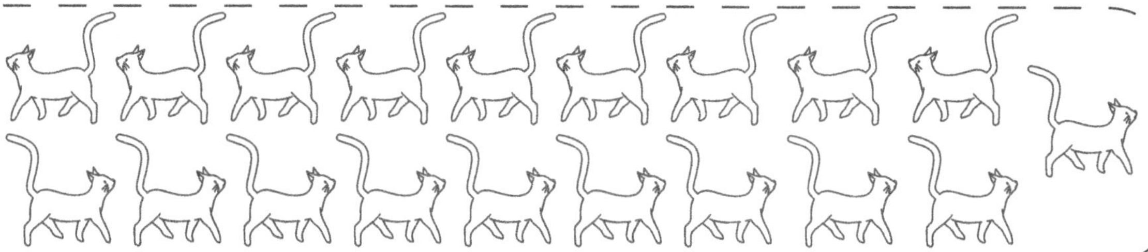

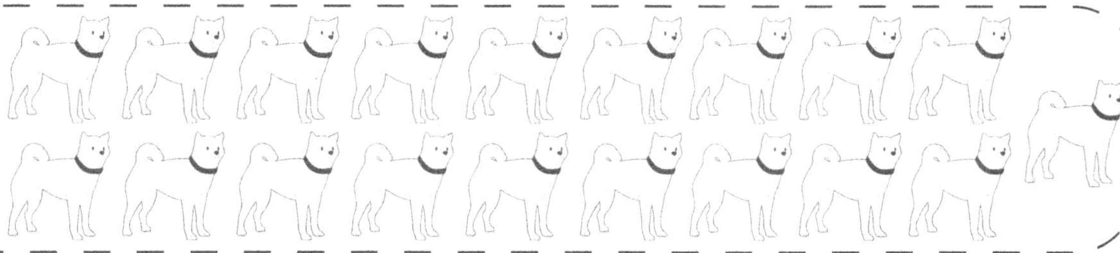

COUNTING GAME

COUNT HOW MANY MANY PICTURES IN THE BOX, THEN WRITE THE NUMBER.

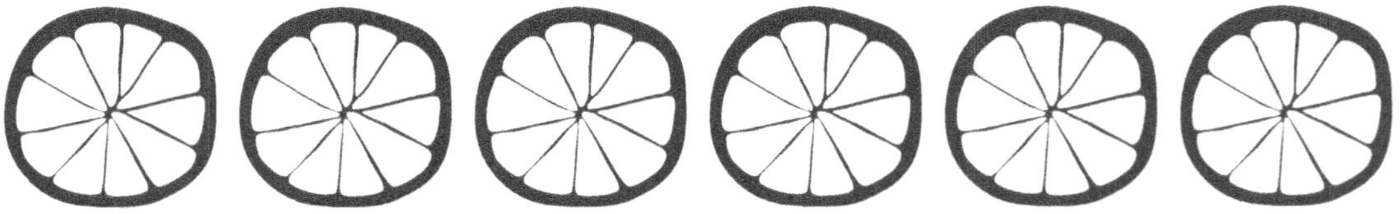

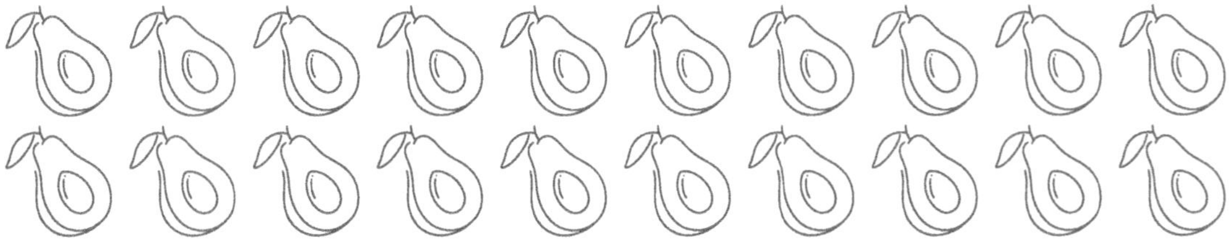

COUNTING GAME

COUNT HOW MANY MANY PICTURES IN THE BOX, THEN WRITE THE NUMBER.

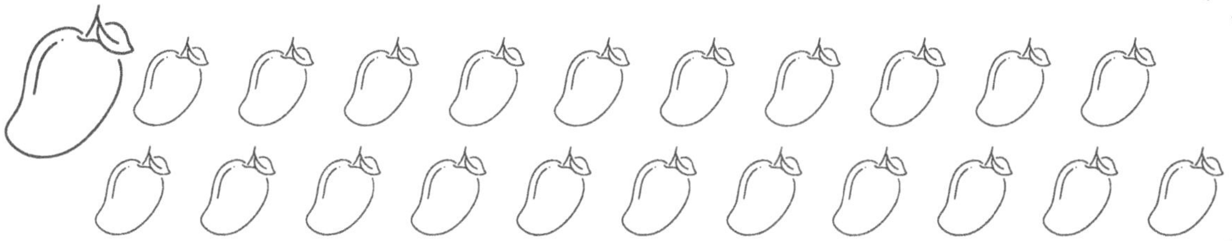

COUNTING GAME

COUNT HOW MANY MANY PICTURES IN THE BOX, THEN WRITE THE NUMBER.

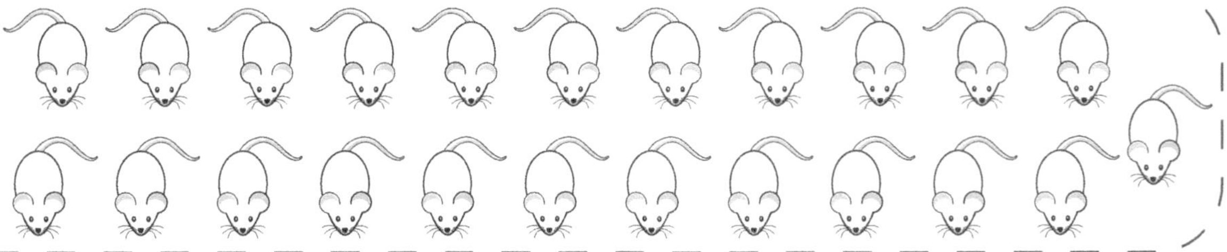

ADDITION

SET A TIMER, SOLVE THE EQUATIONS AND LOG THE TIME AND SCORE BELOW.

TIME: ________ **SCORE:** ________

1) $2 + 1$	2) $1 + 4$	3) $7 + 2$	4) $21 + 4$
5) $1 + 6$	6) $1 + 42$	7) $4 + 2$	8) $5 + 3$
9) $20 + 1$	10) $1 + 8$	11) $44 + 10$	12) $10 + 30$
13) $1 + 8$	14) $60 + 3$	15) $6 + 70$	16) $41 + 1$
17) $1 + 41$	18) $4 + 62$	19) $2 + 50$	20) $5 + 30$

ADDITION

SET A TIMER, SOLVE THE EQUATIONS AND LOG THE TIME AND SCORE BELOW.

TIME: ____________ **SCORE:** ____________

1) $32 + 5 = $ ____

2) $33 + 4 = $ ____

3) $26 + 2 = $ ____

4) $11 + 3 = $ ____

5) $7 + 20 = $ ____

6) $4 + 55 = $ ____

7) $8 + 20 = $ ____

8) $30 + 1 = $ ____

9) $6 + 41 = $ ____

10) $6 + 21 = $ ____

11) $2 + 74 = $ ____

12) $4 + 21 = $ ____

13) $41 + 5 = $ ____

14) $7 + 30 = $ ____

15) $61 + 7 = $ ____

16) $11 + 7 = $ ____

17) $7 + 11 = $ ____

18) $3 + 33 = $ ____

19) $71 + 5 = $ ____

20) $36 + 1 = $ ____

ADDITION

SET A TIMER, SOLVE THE EQUATIONS AND LOG THE TIME AND SCORE BELOW.

TIME: ____________ **SCORE:** ____________

1) $30 + 43$	2) $73 + 10$	3) $40 + 59$	4) $23 + 26$
5) $32 + 11$	6) $13 + 22$	7) $10 + 39$	8) $51 + 12$
9) $60 + 28$	10) $62 + 36$	11) $60 + 10$	12) $43 + 51$
13) $23 + 63$	14) $30 + 59$	15) $22 + 27$	16) $12 + 62$
17) $35 + 20$	18) $43 + 54$	19) $21 + 78$	20) $65 + 20$

ADDITION

SET A TIMER, SOLVE THE EQUATIONS AND LOG THE TIME AND SCORE BELOW.

TIME: ____________ **SCORE:** ____________

1) 72 + 16	2) 73 + 11	3) 38 + 51	4) 15 + 12
5) 33 + 63	6) 60 + 37	7) 39 + 50	8) 83 + 11
9) 33 + 46	10) 60 + 10	11) 34 + 32	12) 21 + 25
13) 23 + 23	14) 12 + 36	15) 42 + 12	16) 21 + 77
17) 33 + 56	18) 42 + 40	19) 22 + 43	20) 54 + 13

ADDITION

SET A TIMER, SOLVE THE EQUATIONS AND LOG THE TIME AND SCORE BELOW.

TIME: ________ **SCORE:** ________

1) $31 + 37 =$

2) $29 + 50 =$

3) $70 + 17 =$

4) $13 + 54 =$

5) $38 + 11 =$

6) $17 + 31 =$

7) $21 + 23 =$

8) $41 + 37 =$

9) $21 + 47 =$

10) $25 + 62 =$

11) $13 + 56 =$

12) $68 + 11 =$

13) $10 + 29 =$

14) $12 + 72 =$

15) $13 + 12 =$

16) $30 + 34 =$

17) $12 + 62 =$

18) $17 + 70 =$

19) $64 + 10 =$

20) $51 + 33 =$

ADDITION

SET A TIMER, SOLVE THE EQUATIONS AND LOG THE TIME AND SCORE BELOW.

TIME: ______________ **SCORE:** ______________

1) $71 + 28 =$ ____

2) $32 + 21 =$ ____

3) $41 + 26 =$ ____

4) $58 + 31 =$ ____

5) $14 + 11 =$ ____

6) $23 + 26 =$ ____

7) $12 + 60 =$ ____

8) $53 + 40 =$ ____

9) $31 + 23 =$ ____

10) $30 + 64 =$ ____

11) $13 + 10 =$ ____

12) $66 + 12 =$ ____

13) $44 + 30 =$ ____

14) $82 + 14 =$ ____

15) $12 + 53 =$ ____

16) $12 + 65 =$ ____

17) $31 + 30 =$ ____

18) $48 + 30 =$ ____

19) $40 + 58 =$ ____

20) $10 + 36 =$ ____

ADDITION

SET A TIMER, SOLVE THE EQUATIONS AND LOG THE TIME AND SCORE BELOW.

TIME: ___________ **SCORE:** ___________

1) 52 + 15 = ___	2) 27 + 40 = ___	3) 46 + 33 = ___	4) 30 + 65 = ___
5) 15 + 33 = ___	6) 13 + 54 = ___	7) 40 + 33 = ___	8) 16 + 11 = ___
9) 43 + 20 = ___	10) 30 + 43 = ___	11) 14 + 81 = ___	12) 86 + 12 = ___
13) 10 + 79 = ___	14) 85 + 10 = ___	15) 40 + 10 = ___	16) 16 + 82 = ___
17) 44 + 34 = ___	18) 11 + 53 = ___	19) 57 + 20 = ___	20) 63 + 12 = ___

ADDITION

SET A TIMER, SOLVE THE EQUATIONS AND LOG THE TIME AND SCORE BELOW.

TIME: ____________ **SCORE:** ____________

1) 67 + 30	2) 69 + 10	3) 31 + 40	4) 71 + 16
5) 11 + 20	6) 41 + 50	7) 13 + 62	8) 49 + 20
9) 21 + 56	10) 21 + 23	11) 31 + 16	12) 55 + 22
13) 11 + 14	14) 26 + 72	15) 53 + 36	16) 20 + 12
17) 70 + 10	18) 10 + 70	19) 30 + 45	20) 30 + 39

ADDITION

SET A TIMER, SOLVE THE EQUATIONS AND LOG THE TIME AND SCORE BELOW.

TIME: ____________ **SCORE:** ____________

1) $42 + 31 =$ ___

2) $37 + 12 =$ ___

3) $20 + 45 =$ ___

4) $17 + 60 =$ ___

5) $54 + 31 =$ ___

6) $32 + 37 =$ ___

7) $11 + 53 =$ ___

8) $10 + 42 =$ ___

9) $22 + 10 =$ ___

10) $22 + 64 =$ ___

11) $56 + 12 =$ ___

12) $16 + 70 =$ ___

13) $20 + 19 =$ ___

14) $14 + 50 =$ ___

15) $24 + 52 =$ ___

16) $42 + 43 =$ ___

17) $16 + 82 =$ ___

18) $30 + 44 =$ ___

19) $10 + 58 =$ ___

20) $79 + 10 =$ ___

ADDITION

SET A TIMER, SOLVE THE EQUATIONS AND LOG THE TIME AND SCORE BELOW.

TIME: ___________ **SCORE:** ___________

1) $20 + 17 =$ ____

2) $55 + 13 =$ ____

3) $52 + 40 =$ ____

4) $17 + 12 =$ ____

5) $60 + 26 =$ ____

6) $10 + 28 =$ ____

7) $14 + 55 =$ ____

8) $28 + 40 =$ ____

9) $23 + 56 =$ ____

10) $60 + 10 =$ ____

11) $20 + 51 =$ ____

12) $17 + 22 =$ ____

13) $65 + 12 =$ ____

14) $12 + 34 =$ ____

15) $55 + 23 =$ ____

16) $36 + 21 =$ ____

17) $58 + 11 =$ ____

18) $53 + 46 =$ ____

19) $23 + 36 =$ ____

20) $14 + 11 =$ ____

SUBTRACTION

SET A TIMER, SOLVE THE EQUATIONS AND LOG THE TIME AND SCORE BELOW.

TIME: ________ **SCORE:** ________

1) $5 - 3$ = ____

2) $4 - 1$ = ____

3) $8 - 1$ = ____

4) $9 - 6$ = ____

5) $7 - 4$ = ____

6) $7 - 3$ = ____

7) $8 - 2$ = ____

8) $9 - 0$ = ____

9) $6 - 1$ = ____

10) $70 - 50$ = ____

11) $8 - 0$ = ____

12) $99 - 40$ = ____

13) $8 - 5$ = ____

14) $8 - 3$ = ____

15) $3 - 2$ = ____

16) $8 - 7$ = ____

17) $9 - 2$ = ____

18) $9 - 8$ = ____

19) $8 - 7$ = ____

20) $9 - 6$ = ____

SUBTRACTION

SET A TIMER, SOLVE THE EQUATIONS AND LOG THE TIME AND SCORE BELOW.

TIME: ____________ **SCORE:** ____________

1) $68 - 7$

2) $88 - 3$

3) $85 - 2$

4) $27 - 4$

5) $43 - 2$

6) $96 - 4$

7) $79 - 5$

8) $98 - 6$

9) $67 - 5$

10) $96 - 50$

11) $62 - 1$

12) $93 - 11$

13) $77 - 4$

14) $79 - 2$

15) $86 - 3$

16) $79 - 1$

17) $33 - 1$

18) $77 - 2$

19) $46 - 10$

20) $98 - 30$

SUBTRACTION

SET A TIMER, SOLVE THE EQUATIONS AND LOG THE TIME AND SCORE BELOW.

TIME: ________ **SCORE:** ________

1) $89 - 33 =$ ____

2) $86 - 12 =$ ____

3) $58 - 44 =$ ____

4) $89 - 60 =$ ____

5) $95 - 12 =$ ____

6) $95 - 84 =$ ____

7) $89 - 30 =$ ____

8) $70 - 40 =$ ____

9) $97 - 16 =$ ____

10) $96 - 31 =$ ____

11) $76 - 64 =$ ____

12) $82 - 21 =$ ____

13) $63 - 42 =$ ____

14) $23 - 10 =$ ____

15) $69 - 54 =$ ____

16) $77 - 20 =$ ____

17) $74 - 31 =$ ____

18) $75 - 32 =$ ____

19) $81 - 20 =$ ____

20) $85 - 53 =$ ____

SUBTRACTION

SET A TIMER, SOLVE THE EQUATIONS AND LOG THE TIME AND SCORE BELOW.

TIME: ____________ **SCORE:** ____________

1) $98 - 63 =$ ____

2) $66 - 25 =$ ____

3) $97 - 30 =$ ____

4) $39 - 14 =$ ____

5) $77 - 55 =$ ____

6) $77 - 22 =$ ____

7) $54 - 13 =$ ____

8) $79 - 16 =$ ____

9) $85 - 33 =$ ____

10) $96 - 60 =$ ____

11) $83 - 61 =$ ____

12) $99 - 67 =$ ____

13) $49 - 35 =$ ____

14) $98 - 86 =$ ____

15) $79 - 50 =$ ____

16) $66 - 10 =$ ____

17) $71 - 10 =$ ____

18) $68 - 57 =$ ____

19) $59 - 38 =$ ____

20) $99 - 23 =$ ____

SUBTRACTION

SET A TIMER, SOLVE THE EQUATIONS AND LOG THE TIME AND SCORE BELOW.

TIME: ____________ **SCORE:** ____________

1) 58 − 42 = ____

2) 77 − 51 = ____

3) 82 − 40 = ____

4) 93 − 40 = ____

5) 77 − 11 = ____

6) 98 − 46 = ____

7) 48 − 20 = ____

8) 62 − 51 = ____

9) 51 − 20 = ____

10) 94 − 11 = ____

11) 87 − 71 = ____

12) 89 − 16 = ____

13) 68 − 32 = ____

14) 99 − 43 = ____

15) 84 − 32 = ____

16) 93 − 80 = ____

17) 86 − 25 = ____

18) 69 − 37 = ____

19) 59 − 40 = ____

20) 87 − 23 = ____

SUBTRACTION

SET A TIMER, SOLVE THE EQUATIONS AND LOG THE TIME AND SCORE BELOW.

TIME: ____________ **SCORE:** ____________

1) $\begin{array}{r} 73 \\ -\ 21 \\ \hline \end{array}$

2) $\begin{array}{r} 95 \\ -\ 61 \\ \hline \end{array}$

3) $\begin{array}{r} 37 \\ -\ 13 \\ \hline \end{array}$

4) $\begin{array}{r} 97 \\ -\ 11 \\ \hline \end{array}$

5) $\begin{array}{r} 80 \\ -\ 40 \\ \hline \end{array}$

6) $\begin{array}{r} 66 \\ -\ 51 \\ \hline \end{array}$

7) $\begin{array}{r} 86 \\ -\ 13 \\ \hline \end{array}$

8) $\begin{array}{r} 41 \\ -\ 10 \\ \hline \end{array}$

9) $\begin{array}{r} 76 \\ -\ 14 \\ \hline \end{array}$

10) $\begin{array}{r} 73 \\ -\ 52 \\ \hline \end{array}$

11) $\begin{array}{r} 69 \\ -\ 57 \\ \hline \end{array}$

12) $\begin{array}{r} 61 \\ -\ 10 \\ \hline \end{array}$

13) $\begin{array}{r} 87 \\ -\ 72 \\ \hline \end{array}$

14) $\begin{array}{r} 87 \\ -\ 32 \\ \hline \end{array}$

15) $\begin{array}{r} 85 \\ -\ 41 \\ \hline \end{array}$

16) $\begin{array}{r} 99 \\ -\ 47 \\ \hline \end{array}$

17) $\begin{array}{r} 59 \\ -\ 28 \\ \hline \end{array}$

18) $\begin{array}{r} 68 \\ -\ 56 \\ \hline \end{array}$

19) $\begin{array}{r} 79 \\ -\ 33 \\ \hline \end{array}$

20) $\begin{array}{r} 93 \\ -\ 41 \\ \hline \end{array}$

SUBTRACTION

SET A TIMER, SOLVE THE EQUATIONS AND LOG THE TIME AND SCORE BELOW.

TIME: ____________ **SCORE:** ____________

1) 29 − 15	2) 41 − 20	3) 85 − 12	4) 29 − 16
5) 49 − 18	6) 97 − 52	7) 76 − 54	8) 85 − 42
9) 65 − 40	10) 25 − 10	11) 96 − 32	12) 39 − 13
13) 63 − 10	14) 85 − 72	15) 78 − 63	16) 99 − 72
17) 78 − 11	18) 98 − 71	19) 95 − 21	20) 73 − 22

SUBTRACTION

SET A TIMER, SOLVE THE EQUATIONS AND LOG THE TIME AND SCORE BELOW.

TIME: ____________ **SCORE:** ____________

1) $48 - 23$	2) $96 - 82$	3) $99 - 75$	4) $89 - 53$
5) $87 - 60$	6) $88 - 12$	7) $99 - 73$	8) $79 - 31$
9) $39 - 23$	10) $98 - 70$	11) $54 - 23$	12) $99 - 64$
13) $56 - 34$	14) $30 - 10$	15) $70 - 50$	16) $65 - 42$
17) $74 - 12$	18) $95 - 51$	19) $43 - 22$	20) $45 - 24$

SUBTRACTION

SET A TIMER, SOLVE THE EQUATIONS AND LOG THE TIME AND SCORE BELOW.

TIME: ________ **SCORE:** ________

1) $95 - 70 =$ ____

2) $59 - 44 =$ ____

3) $85 - 12 =$ ____

4) $37 - 25 =$ ____

5) $98 - 11 =$ ____

6) $89 - 65 =$ ____

7) $98 - 36 =$ ____

8) $90 - 10 =$ ____

9) $77 - 63 =$ ____

10) $97 - 45 =$ ____

11) $93 - 82 =$ ____

12) $69 - 36 =$ ____

13) $75 - 60 =$ ____

14) $34 - 11 =$ ____

15) $35 - 21 =$ ____

16) $87 - 12 =$ ____

17) $97 - 50 =$ ____

18) $99 - 61 =$ ____

19) $84 - 21 =$ ____

20) $32 - 20 =$ ____

SUBTRACTION

SET A TIMER, SOLVE THE EQUATIONS AND LOG THE TIME AND SCORE BELOW.

TIME: __________ **SCORE:** __________

1) $86 - 51$ = ____

2) $97 - 72$ = ____

3) $55 - 11$ = ____

4) $89 - 73$ = ____

5) $54 - 42$ = ____

6) $96 - 22$ = ____

7) $68 - 54$ = ____

8) $66 - 34$ = ____

9) $49 - 38$ = ____

10) $88 - 62$ = ____

11) $45 - 22$ = ____

12) $78 - 55$ = ____

13) $98 - 34$ = ____

14) $69 - 35$ = ____

15) $87 - 40$ = ____

16) $97 - 51$ = ____

17) $77 - 66$ = ____

18) $98 - 40$ = ____

19) $94 - 43$ = ____

20) $43 - 10$ = ____

TIMES TABLE

1x

1 x 1 = 1
2 x 1 = 2
3 x 1 = 3
4 x 1 = 4
5 x 1 = 5
6 x 1 = 6
7 x 1 = 7
8 x 1 = 8
9 x 1 = 9
10 x 1 = 10
11 x 1 = 11
12 x 1 = 12

2x

1 x 2 = 2
2 x 2 = 4
3 x 2 = 6
4 x 2 = 8
5 x 2 = 10
6 x 2 = 12
7 x 2 = 14
8 x 2 = 16
9 x 2 = 18
10 x 2 = 20
11 x 2 = 22
12 x 2 = 24

3x

1 x 3 = 3
2 x 3 = 6
3 x 3 = 9
4 x 3 = 12
5 x 3 = 15
6 x 3 = 18
7 x 3 = 21
8 x 3 = 24
9 x 3 = 27
10 x 3 = 30
11 x 3 = 33
12 x 3 = 36

4x

1 x 4 = 4
2 x 4 = 8
3 x 4 = 12
4 x 4 = 16
5 x 4 = 20
6 x 4 = 24
7 x 4 = 28
8 x 4 = 32
9 x 4 = 36
10 x 4 = 40
11 x 4 = 44
12 x 4 = 48

5x

1 x 5 = 5
2 x 5 = 10
3 x 5 = 15
4 x 5 = 20
5 x 5 = 25
6 x 5 = 30
7 x 5 = 35
8 x 5 = 40
9 x 5 = 45
10 x 5 = 50
11 x 5 = 55
12 x 5 = 60

6x

1 x 6 = 6
2 x 6 = 12
3 x 6 = 18
4 x 6 = 24
5 x 6 = 30
6 x 6 = 36
7 x 6 = 42
8 x 6 = 48
9 x 6 = 54
10 x 6 = 60
11 x 6 = 66
12 x 6 = 72

7x

1 x 7 = 7
2 x 7 = 14
3 x 7 = 21
4 x 7 = 28
5 x 7 = 35
6 x 7 = 42
7 x 7 = 49
8 x 7 = 56
9 x 7 = 63
10 x 7 = 70
11 x 7 = 77
12 x 7 = 84

8x

1 x 8 = 8
2 x 8 = 16
3 x 8 = 24
4 x 8 = 32
5 x 8 = 40
6 x 8 = 48
7 x 8 = 56
8 x 8 = 64
9 x 8 = 72
10 x 8 = 80
11 x 8 = 88
12 x 8 = 96

9x

1 x 9 = 9
2 x 9 = 18
3 x 9 = 27
4 x 9 = 36
5 x 9 = 45
6 x 9 = 54
7 x 9 = 63
8 x 9 = 72
9 x 9 = 81
10 x 9 = 90
11 x 9 = 99
12 x 9 = 108

10x

1 x 10 = 10
2 x 10 = 20
3 x 10 = 30
4 x 10 = 40
5 x 10 = 50
6 x 10 = 60
7 x 10 = 70
8 x 10 = 80
9 x 10 = 90
10 x 10 = 100
11 x 10 = 110
12 x 10 = 120

11x

1 x 11 = 11
2 x 11 = 22
3 x 11 = 33
4 x 11 = 44
5 x 11 = 55
6 x 11 = 66
7 x 11 = 77
8 x 11 = 88
9 x 11 = 99
10 x 11 = 110
11 x 11 = 121
12 x 11 = 132

12x

1 x 12 = 12
2 x 12 = 24
3 x 12 = 36
4 x 12 = 48
5 x 12 = 60
6 x 12 = 72
7 x 12 = 84
8 x 12 = 96
9 x 12 = 108
10 x 12 = 120
11 x 12 = 132
12 x 12 = 144

MULTIPLICATION

SET A TIMER, SOLVE THE EQUATIONS AND LOG THE TIME AND SCORE BELOW.

TIME: ________ **SCORE:** ________

1) 10×3

2) 7×1

3) 5×5

4) 6×7

5) 12×12

6) 14×15

7) 6×10

8) 10×1

9) 15×4

10) 4×7

11) 2×9

12) 2×2

13) 7×13

14) 13×3

15) 10×9

16) 5×3

17) 12×11

18) 6×8

19) 13×12

20) 14×2

MULTIPLICATION

SET A TIMER, SOLVE THE EQUATIONS AND LOG THE TIME AND SCORE BELOW.

TIME: ________ **SCORE:** ________

1) 1×10

2) 5×6

3) 6×3

4) 2×2

5) 11×7

6) 14×11

7) 10×1

8) 11×13

9) 4×9

10) 14×15

11) 5×8

12) 4×11

13) 10×9

14) 3×9

15) 1×11

16) 12×6

17) 8×7

18) 6×12

19) 14×6

20) 3×4

MULTIPLICATION

SET A TIMER, SOLVE THE EQUATIONS AND LOG THE TIME AND SCORE BELOW.

TIME: ________ **SCORE:** ________

1) 7×9

2) 15×12

3) 1×9

4) 6×10

5) 15×3

6) 11×9

7) 1×3

8) 11×14

9) 15×7

10) 8×5

11) 4×14

12) 6×4

13) 15×5

14) 2×7

15) 6×12

16) 15×11

17) 7×12

18) 12×12

19) 7×8

20) 6×1

MULTIPLICATION

SET A TIMER, SOLVE THE EQUATIONS AND LOG THE TIME AND SCORE BELOW.

TIME: ________ **SCORE:** ________

1) 11×6

2) 3×6

3) 3×12

4) 2×3

5) 6×15

6) 14×13

7) 13×15

8) 8×5

9) 6×2

10) 11×11

11) 9×3

12) 3×7

13) 6×10

14) 4×2

15) 15×13

16) 10×5

17) 1×12

18) 1×15

19) 13×8

20) 13×1

MULTIPLICATION

SET A TIMER, SOLVE THE EQUATIONS AND LOG THE TIME AND SCORE BELOW.

TIME: ________ **SCORE:** ________

1) 7×13

2) 2×10

3) 15×12

4) 11×15

5) 12×14

6) 15×14

7) 14×9

8) 11×4

9) 8×10

10) 1×4

11) 15×9

12) 14×6

13) 5×3

14) 8×15

15) 12×6

16) 13×7

17) 13×4

18) 6×3

19) 7×7

20) 11×1

MULTIPLICATION

SET A TIMER, SOLVE THE EQUATIONS AND LOG THE TIME AND SCORE BELOW.

TIME: ________ **SCORE:** ________

1) 2×13

2) 11×13

3) 4×5

4) 5×15

5) 13×5

6) 3×2

7) 13×2

8) 8×13

9) 4×6

10) 12×3

11) 15×8

12) 11×5

13) 6×15

14) 15×7

15) 10×9

16) 15×3

17) 9×1

18) 3×1

19) 5×4

20) 9×4

MULTIPLICATION

SET A TIMER, SOLVE THE EQUATIONS AND LOG THE TIME AND SCORE BELOW.

TIME: ________ **SCORE:** ________

1) 12×3

2) 5×3

3) 14×7

4) 3×15

5) 2×7

6) 13×1

7) 11×2

8) 7×7

9) 1×11

10) 6×10

11) 6×12

12) 10×12

13) 4×12

14) 10×15

15) 14×8

16) 1×4

17) 7×11

18) 12×9

19) 9×13

20) 9×14

MULTIPLICATION

SET A TIMER, SOLVE THE EQUATIONS AND LOG THE TIME AND SCORE BELOW.

TIME: ________ **SCORE:** ________

1) 5×14

2) 4×13

3) 5×1

4) 14×1

5) 6×8

6) 2×4

7) 15×10

8) 3×6

9) 7×5

10) 3×11

11) 1×7

12) 2×2

13) 8×5

14) 1×1

15) 10×3

16) 15×9

17) 8×3

18) 4×1

19) 1×8

20) 15×13

MULTIPLICATION

SET A TIMER, SOLVE THE EQUATIONS AND LOG THE TIME AND SCORE BELOW.

TIME: ________ **SCORE:** ________

1) 6×14	2) 5×12	3) 14×7	4) 11×5
5) 11×2	6) 13×4	7) 6×11	8) 4×4
9) 13×7	10) 14×8	11) 7×14	12) 12×5
13) 6×7	14) 9×13	15) 3×3	16) 4×7
17) 7×8	18) 1×1	19) 12×15	20) 13×15

MULTIPLICATION

SET A TIMER, SOLVE THE EQUATIONS AND LOG THE TIME AND SCORE BELOW.

TIME: __________ **SCORE:** __________

1) $10 \times 12 =$ ____

2) $1 \times 12 =$ ____

3) $13 \times 1 =$ ____

4) $7 \times 1 =$ ____

5) $5 \times 13 =$ ____

6) $5 \times 11 =$ ____

7) $1 \times 5 =$ ____

8) $7 \times 12 =$ ____

9) $6 \times 5 =$ ____

10) $5 \times 6 =$ ____

11) $6 \times 3 =$ ____

12) $11 \times 2 =$ ____

13) $14 \times 9 =$ ____

14) $9 \times 8 =$ ____

15) $14 \times 1 =$ ____

16) $2 \times 6 =$ ____

17) $9 \times 11 =$ ____

18) $6 \times 13 =$ ____

19) $10 \times 2 =$ ____

20) $12 \times 8 =$ ____

DIVISION

SET A TIMER, SOLVE THE EQUATIONS AND LOG THE TIME AND SCORE BELOW.

TIME: ____________ **SCORE:** ____________

1) $77 \div 11$

2) $85 \div 17$

3) $46 \div 23$

4) $30 \div 15$

5) $45 \div 15$

6) $99 \div 33$

7) $60 \div 12$

8) $86 \div 43$

9) $78 \div 13$

10) $88 \div 11$

11) $56 \div 14$

12) $99 \div 11$

13) $63 \div 21$

14) $66 \div 11$

15) $84 \div 42$

16) $40 \div 10$

17) $75 \div 15$

18) $66 \div 33$

19) $92 \div 46$

20) $22 \div 11$

DIVISION

SET A TIMER, SOLVE THE EQUATIONS AND LOG THE TIME AND SCORE BELOW.

TIME: ___________ **SCORE:** ___________

1) $80 \div 16 =$ ____

2) $54 \div 18 =$ ____

3) $42 \div 14 =$ ____

4) $72 \div 12 =$ ____

5) $48 \div 16 =$ ____

6) $54 \div 27 =$ ____

7) $96 \div 48 =$ ____

8) $50 \div 25 =$ ____

9) $45 \div 15 =$ ____

10) $57 \div 19 =$ ____

11) $24 \div 12 =$ ____

12) $20 \div 10 =$ ____

13) $46 \div 23 =$ ____

14) $51 \div 17 =$ ____

15) $76 \div 38 =$ ____

16) $50 \div 10 =$ ____

17) $99 \div 33 =$ ____

18) $30 \div 10 =$ ____

19) $44 \div 11 =$ ____

20) $70 \div 10 =$ ____

DIVISION

SET A TIMER, SOLVE THE EQUATIONS AND LOG THE TIME AND SCORE BELOW.

TIME: ____________ **SCORE:** ____________

1) $84 \div 28 =$ ____

2) $22 \div 11 =$ ____

3) $66 \div 22 =$ ____

4) $60 \div 30 =$ ____

5) $32 \div 16 =$ ____

6) $90 \div 10 =$ ____

7) $86 \div 43 =$ ____

8) $44 \div 11 =$ ____

9) $76 \div 38 =$ ____

10) $44 \div 22 =$ ____

11) $80 \div 20 =$ ____

12) $72 \div 24 =$ ____

13) $48 \div 16 =$ ____

14) $80 \div 16 =$ ____

15) $20 \div 10 =$ ____

16) $70 \div 14 =$ ____

17) $30 \div 10 =$ ____

18) $36 \div 18 =$ ____

19) $74 \div 37 =$ ____

20) $80 \div 10 =$ ____

DIVISION

SET A TIMER, SOLVE THE EQUATIONS AND LOG THE TIME AND SCORE BELOW.

TIME: ________ **SCORE:** ________

1) $98 \div 14 =$ ____

2) $24 \div 12 =$ ____

3) $64 \div 16 =$ ____

4) $64 \div 32 =$ ____

5) $94 \div 47 =$ ____

6) $86 \div 43 =$ ____

7) $82 \div 41 =$ ____

8) $84 \div 28 =$ ____

9) $70 \div 14 =$ ____

10) $77 \div 11 =$ ____

11) $72 \div 24 =$ ____

12) $30 \div 15 =$ ____

13) $33 \div 11 =$ ____

14) $90 \div 45 =$ ____

15) $42 \div 21 =$ ____

16) $76 \div 38 =$ ____

17) $78 \div 13 =$ ____

18) $52 \div 13 =$ ____

19) $66 \div 11 =$ ____

20) $80 \div 40 =$ ____

DIVISION

SET A TIMER, SOLVE THE EQUATIONS AND LOG THE TIME AND SCORE BELOW.

TIME: ____________ **SCORE:** ____________

1) $52 \div 26$	2) $99 \div 11$	3) $40 \div 10$	4) $77 \div 11$
5) $91 \div 13$	6) $20 \div 10$	7) $78 \div 26$	8) $80 \div 16$
9) $76 \div 19$	10) $44 \div 22$	11) $94 \div 47$	12) $87 \div 29$
13) $72 \div 18$	14) $95 \div 19$	15) $42 \div 14$	16) $86 \div 43$
17) $64 \div 16$	18) $60 \div 15$	19) $72 \div 24$	20) $63 \div 21$

DIVISION

SET A TIMER, SOLVE THE EQUATIONS AND LOG THE TIME AND SCORE BELOW.

TIME: ______ **SCORE:** ______

1) $60 \div 20$

2) $75 \div 25$

3) $92 \div 23$

4) $88 \div 44$

5) $36 \div 12$

6) $78 \div 13$

7) $70 \div 14$

8) $44 \div 22$

9) $86 \div 43$

10) $66 \div 33$

11) $57 \div 19$

12) $39 \div 13$

13) $44 \div 11$

14) $32 \div 16$

15) $78 \div 26$

16) $76 \div 19$

17) $80 \div 16$

18) $66 \div 11$

19) $54 \div 18$

20) $66 \div 22$

DIVISION

SET A TIMER, SOLVE THE EQUATIONS AND LOG THE TIME AND SCORE BELOW.

TIME: ________ **SCORE:** ________

1) $60 \div 15$	2) $54 \div 18$	3) $76 \div 38$	4) $45 \div 15$
5) $93 \div 31$	6) $48 \div 12$	7) $96 \div 24$	8) $69 \div 23$
9) $90 \div 18$	10) $68 \div 17$	11) $32 \div 16$	12) $98 \div 14$
13) $60 \div 10$	14) $51 \div 17$	15) $78 \div 39$	16) $22 \div 11$
17) $84 \div 28$	18) $86 \div 43$	19) $55 \div 11$	20) $95 \div 19$

DIVISION

SET A TIMER, SOLVE THE EQUATIONS AND LOG THE TIME AND SCORE BELOW.

TIME: ______________ **SCORE:** ______________

1) $84 \div 21$

2) $88 \div 11$

3) $40 \div 10$

4) $32 \div 16$

5) $93 \div 31$

6) $72 \div 24$

7) $96 \div 48$

8) $78 \div 39$

9) $42 \div 14$

10) $66 \div 11$

11) $65 \div 13$

12) $45 \div 15$

13) $54 \div 27$

14) $56 \div 28$

15) $72 \div 18$

16) $92 \div 46$

17) $92 \div 23$

18) $50 \div 25$

19) $64 \div 32$

20) $76 \div 38$

DIVISION

SET A TIMER, SOLVE THE EQUATIONS AND LOG THE TIME AND SCORE BELOW.

TIME: ________ **SCORE:** ________

1) $22 \div 11 =$ ____

2) $80 \div 20 =$ ____

3) $80 \div 10 =$ ____

4) $20 \div 10 =$ ____

5) $96 \div 16 =$ ____

6) $70 \div 10 =$ ____

7) $30 \div 10 =$ ____

8) $54 \div 18 =$ ____

9) $68 \div 17 =$ ____

10) $60 \div 12 =$ ____

11) $42 \div 21 =$ ____

12) $40 \div 20 =$ ____

13) $52 \div 13 =$ ____

14) $92 \div 23 =$ ____

15) $78 \div 26 =$ ____

16) $92 \div 46 =$ ____

17) $57 \div 19 =$ ____

18) $39 \div 13 =$ ____

19) $26 \div 13 =$ ____

20) $42 \div 14 =$ ____

DIVISION

SET A TIMER, SOLVE THE EQUATIONS AND LOG THE TIME AND SCORE BELOW.

TIME: ________ **SCORE:** ________

1) $93 \div 31$

2) $88 \div 22$

3) $85 \div 17$

4) $22 \div 11$

5) $24 \div 12$

6) $30 \div 15$

7) $81 \div 27$

8) $48 \div 24$

9) $84 \div 21$

10) $65 \div 13$

11) $62 \div 31$

12) $55 \div 11$

13) $76 \div 38$

14) $98 \div 49$

15) $40 \div 10$

16) $88 \div 11$

17) $86 \div 43$

18) $60 \div 15$

19) $36 \div 18$

20) $78 \div 26$

Hope you enjoyed this book!
For more books like this one check out
My website
www.authorcjordan.com

Sign up for our newsletter to get updates on new releases and Your Free Downloadable Coloring Book!

Made in United States
Cleveland, OH
05 November 2026

47672907R00056